AMAZON ECHO SHOW

The Ultimate Guide for Complete Beginners On How to Setup Your Amazon Echo Show in Few Minutes.

I0492701

BY

CHARLES S. MILLS

Copyright©2018

COPYRIGHT

Charles S. Mills

TABLE OF CONTENT

CHAPTER 1

INTRODUCTION

The Amazon Echo Show which is more than just a novelty device is one of the world most popular android device that can be both seen and heard and it comes with a 7-inch touchscreen. With over 3 million users who owns it, and enjoys using the amazing features it has in it to the fullest.

Apart from the popular use of Amazon Echo Show device as an E-reader, it can also perform other functions you can think of including Amazon video content, video flash briefings, music lyrics, weather forecast and news update. Everything you have ever dying to know about your Echo

Show including what is the device, how to control a smart home, detailing what skills does it have, troubleshooting common problems, setup the connected device, and much more. The user interface is a friendly kind and it has a collection of enjoyable stuff in it.

This guide will show you how you can setup your Amazon Echo Show, all you have to do to achieve and get acquitted with the Amazon Echo Show device is carefully follow it step by step as they are instructed on this guide. Also you can as well perform wonders with the Amazon Echo Show features it has in it with proper meditations of each step provided here, but if you find any issue very difficult, there is no needs to worry because I got you cover with this guide.

Globally millions of people haven't been able to use and explore wonders with the amazing features it has, but this book gives the breakdown of all the answers to any problem you might face.

Thankfully each steps are very easy and simple to follow when you go through this guide and reading it, hopefully you have gained a good understanding and being enlighten of the Echo Show with proper meditations and how you can explore and make use of its amazing features more interesting and enjoyable. That even a beginner can master it in a few minutes and unleash its full potential.

CHAPTER 2

AMAZON ECHO SHOW

WHAT IS IT?

The Echo Show is a powerful and handy device that can now be both seen and heard and it comes with a touchscreen. The headlining features of the device, currently only available to Amazon Prime members in the U.S., is a 7-inch touchscreen that can perform anything you can think of including Amazon video content, video flash briefings, music lyrics, weather forecast and news update.

Echo Show is effectively what you get when you took an Alexa-integrated Fire tablet, place a powerful speaker on the bottom then frame it with glossy black plastic. The expecting result would be a rather monolithic look. The Echo Show for such a small thing it looks very imposing and interesting, and has split the opinion of visitors to my house 50:50 on whether the Echo Show is attractive or not. If place on a shelf or in a corner, it can easily blend with the surroundings. However, when it's not actively doing something, it can perform like a smart internet connected version of those digital photo frames that went out of fashion quickly. You can pick a rotating catalogue of Amazon's images, just a single picture or a whole album of pictures from your Amazon photo library to show on the Echo Show 7-inch touchscreen.

The Echo Show offers every other amazing features of any Echo device including eight microphones, room-filling sound, and the ability to be heard from a noisy room. The Amazon Echo Show takes away the Alexa voice assistant and squeezes it into a cross between a small TV, digital photo frame and smart speaker for something that is more than just an interesting novelty.

If you have just bought an Echo Show and want to know the best way to leverage it right away, you are advice to have a copy of this guide. Everything you have ever dying to know about your Echo Show including what is the device, how to control a smart home, detailing what skills does Echo Show possess, troubleshooting common problems, setup the connected device, and much more.

CHAPTER 3

UNBOX AND GET YOUR AMAZON ECHO SHOW SETUP

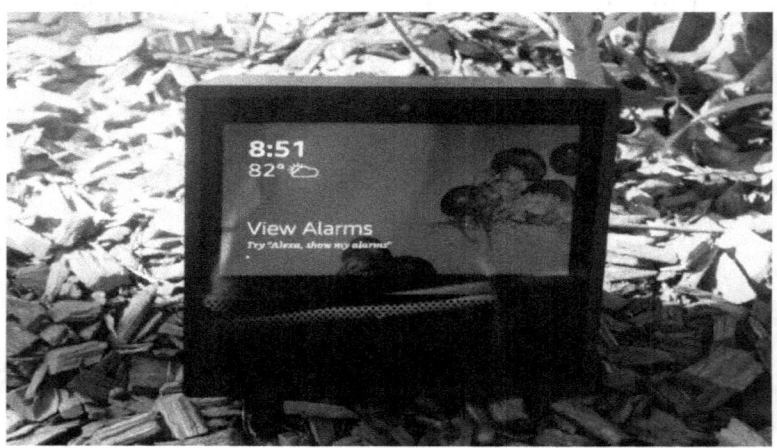

After unboxing your Amazon Echo Show, you are going to find the listed items included inside:

→ The Amazon Echo Show (white or black version)

→ A getting started pamphlet

→ A small guide with some basic Alexa commands in it

→ An adapter and a power cord

The most necessary thing to do before you plug in the Echo Show is downloading the companion Amazon Echo app for your Android or IOS device. Technically you can make use of the Echo Show without the app, as it's a good way to communicate with the device and get a number of different settings customize.

Meanwhile, to enable you use the device you should follow the steps below, after plugging in the Echo Show;

→ You will need to choose a language

→ Input Wi-Fi credentials

→ Enter Amazon account details using the touchscreen.

Since the Echo Show can only be bought by a prime member, to use the device an account will be required.

Thanks for the awareness to the 2-inch stereo speakers that provide very well the best sound profile of any Echo at large.

Now, to get started the most basic first step to take is;

→ Move to the app then link all applicable music services.

→ Unveil the Amazon Alexa app and move to the side menu.

→ Then choose **Music, Video, & Books**.

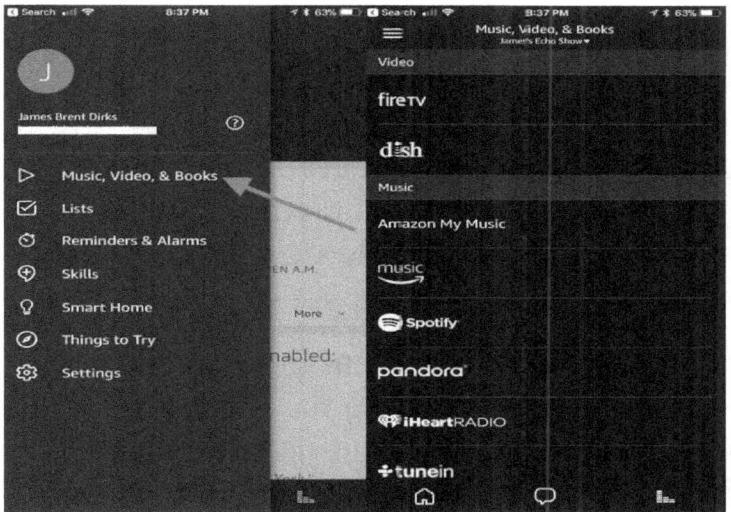

It is able to choose between some different music options to play on the Echo Show, as well as music from Amazon. The Amazon music streaming service, Spotify, TuneIn, iHeartRadio, SiriusXM, Pandora.

Options like Amazon Music and TuneIn does not require a subscription or any login details. But in order to use the Spotify and SiriusXM you have to be a subscriber.

However, in a fun touch especially for those close karaoke singers, lots of songs from Amazon Music will automatically display lyrics and album art on the screen. If you choose to switch it on or off, all you have to do is just say "**Alexa, turn off/on lyrics**".

You will see three different buttons on the top of the Echo Show. On the left side is a **Do Not Disturb** button that can switch off both the microphone and the camera. While the other two buttons are the volume control buttons that will enable you control the speakers without giving a command to Alexa.

Using Bluetooth, you can as well connect the Echo Show to another speaker for audio output. To enable you get this done, just say **"Alexa, go to settings"** or you can find the settings option by scrolling down on the screen. Then select **Bluetooth.** Ensure that the Bluetooth speaker you wish to use is in pairing mode, which will appear on the list of available devices. Select the right Bluetooth from the list of available devices and then follow the instructions as given.

HOUSEHOLD PROFILE

In a case whereby more than one person in your house desire to manage their own music library including other features, the Echo Show has the ability and can support

multiple users and this process is called
Household Profiles.

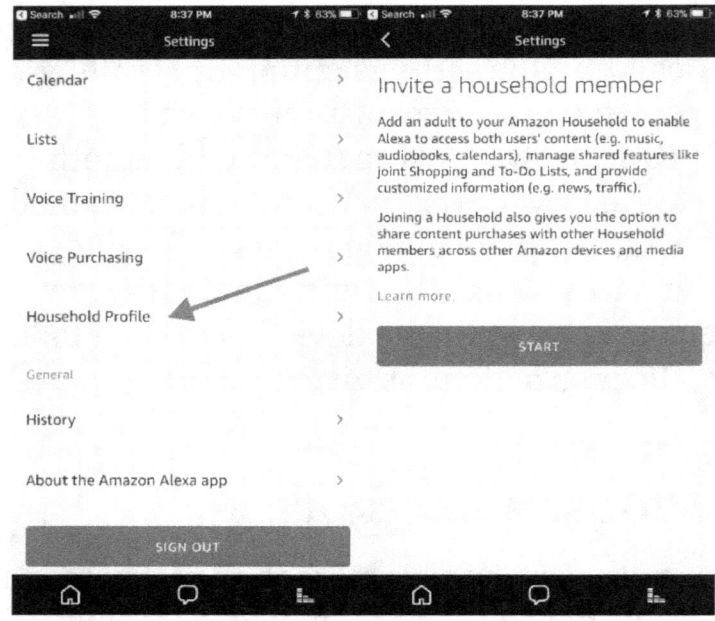

Ensure the presence of the person that you want to add. In the app, move to the menu sidebar and then select;

→ Settings

→ Accounts

→ Household Profiles

After that you should then follow the instructions on the screen.

Immediately added, the other person will be able to view your Prime Photos on the Echo Show. Hence, by doing so you also grant them permission to make purchases using credit cards associated with the Amazon account. However, they can also access customized traffic and news, to-do list, their own music, and shopping list.

Once multiple accounts are set, you should just say **"Alexa, what account is this?"** and **"Alexa, switch accounts"** if it's been needed.

If need be to remove another account, head to;

→ Settings

→ Accounts

→ In an Amazon household with (name).

Then click **Remove** next to the particular user you are trying to remove, you will need to click **Remove** again to confirm your request. Hit **Leave** if you want to remove yourself.

However, get it noted that if another account is removed, it will be unable to add it back to any other household for six months.

CHAPTER 4

GET YOUR AMAZON ECHO SHOW ACQUITTED TO NEW SKILLS

Once you've got your Echo Show setup, one of the most basic and important thing you should know about are skills. Skills give power to the Echo Show and all of the other Alexa enabled devices. A good way to think about skills is that each one is a voice controllable app.

Presently, there are more than 15,000 different skills you can select from. Some skills are not so smart, but there are number of truly useful skills available. Most Echo Show skills take the merit of the touchscreen, but some are still designed primarily for the other audio only in Echo devices. Hence, the number of skills that make use of the Echo Show continually grows.

Currently, there isn't a way know yet if a particular skill is optimized for the screen of the device. Depending on hope that as the number of skills increases change will occur.

WAYS YOU CAN SEARCH AND ENABLE SKILLS

Unfortunately, as of now, there are a few different ways to find and enable skills for your Echo Show. You can also check out all the available options on the Amazon Skills Portal, luck fully the most convenient way is using companion app.

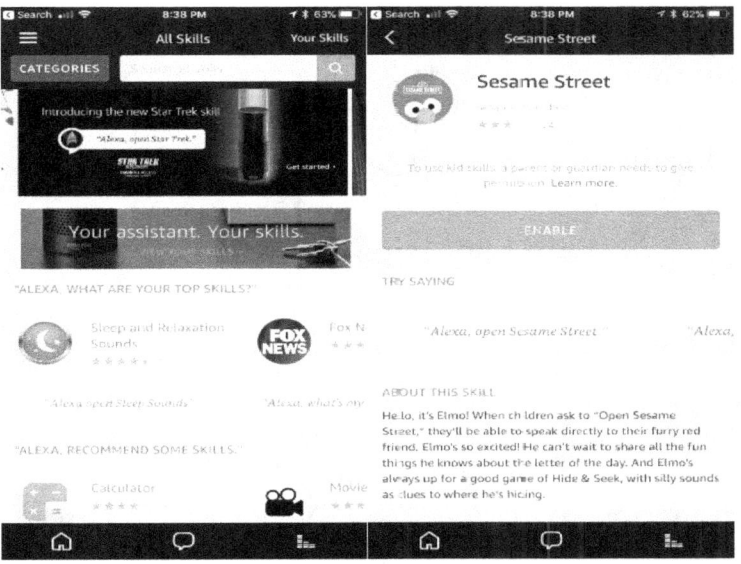

From the side menu bar, click **Skills.** You are able to have a view of a number of different curated list given that show new, trending and recommended skills. To see all the available skills, you should just click on **Categories** next to the search bar.

On each skill page, you will be able to see some specific voice commands, customer reviews, option to receive support regarding the skill, and a skill description. Tap on **Enable** button once you find an amazing skill. Some option command you to log-in to a companion service while others are already set. It's easier if you know the name of the particular skill, all you have to do is just say **"Alexa, enable (skill name)".**

ECHO SHOW SKILL CATEGORIES

Below are all the categories of skills you can choose from:

    a. Lifestyle

    b. Food and Drink

    c. News

    d. Local

e. Social

f. Sports

g. Novelty and Humor

h. Newest Arrivals

i. Connected Cars

j. Weather

k. Shopping

l. Business and Finance

m. Health and Fitness

n. Education and Reference

o. Movies and TV

p. Games, Trivia and Accessories

q. Music and Audio

r. Productivity

s. Utilities

t. Travel and Transportation

u. Smart Home

CHAPTER 5

USING AMAZON ECHO SHOW TO CONTROL YOUR SMART HOME

The Echo Show can complete a huge number of different tasks; most times it currently has advantage in controlling a smart home. The ecosystem of devices that can pair with Alexa is huge and is getting larger day by day.

The convenient way to know if a smart home device can pair with an Echo Show is to view on Amazon's Echo Smart Home page.

INCLUDING AND EXCLUDING A SMART HOME DEVICE

If your device can pair with Echo Show, you should ensure to download the manufacturer's companion app as well.

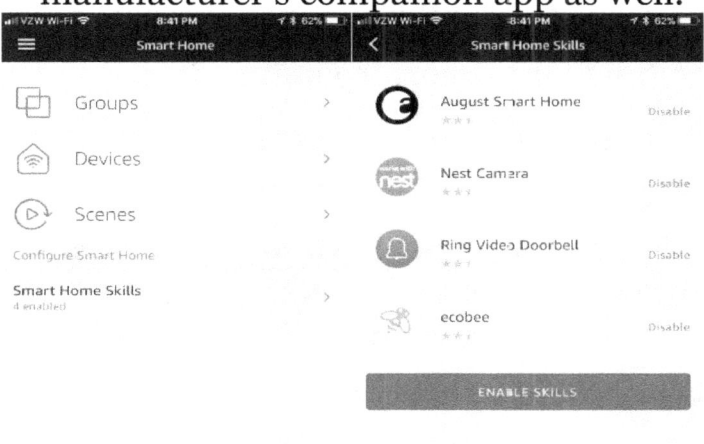

Once that is done, unveil the Alexa app and then from the side menu choose:

→ **Smart Home.**

→ Select **Devices.**

→ After that open the **Alexa Smart Home Store**

Then you should select **Enable,** after searching for the corresponding device.

Note that, it may pop a process to link your device with the service.

Lastly, Alexa will need to find any devices that can be set successful through the app's menu by choosing:

→**Smart Home**

→ **Devices**

→ **Discover**

Also you can as well say **"Alexa, discover devices".** All available devices in the app can be seen or you can just say Alexa how many devices were found. In the discovering process there are two big exceptions. In case you own any Philips Hue or Belkin WeMo devices light with the real version 1 bridge that is circle shaped. To enable a skill won't be needed, as when discovering them they should be found automatically. With the hue bridge, ensure to tap on the button on top before beginning the discovering process.

If at any time need be to remove a smart home device from the Echo Show. On the side menu move to the **Smart Home** section, after choosing **Devices,** you will need to hit **Forget** on each device that you want to remove.

Note, it's also necessary to get rid of the device from the manufacturer's companion app as well.

Simple voice command can be used to control its processes, once the device has been connected.

CONNECTING ECHO SHOW WITH CAMERAS

A good deal for the Echo Show is that it has the ability to connect a number of different smart home cameras. As of now, cameras from famous manufacturers like Nest, Vivint, EZviz, Ring, Arlo, August, Amcrest, Logitech, and IC Realtime are good to pair.

It is possible to add a compatible camera just like any other smart home device by applying the method above.

You should ensure to recall the name of your camera, if you do then you can just say **"Alexa, show (camera name)"** and immediately the screen will play a live video feed with audio. When you are through looking at the camera feed, you should then say **"Alexa, stop"**, **"Alexa,**

hide (camera name)", or you can even say **"Alexa, go home".**

CREATING GROUPS AND DISCOVERING SCENES

If you install more than one smart home device, it's a great idea to create some groups that will make it to appear much easier using a voice command to control your home. Instead of wanting to tell Alexa to adjust each device one by one, for instance, you can create a group named "sitting room". Then you can just say, **"Alexa, switch off the sitting room lights"** and finally everything will be done easily and quickly.

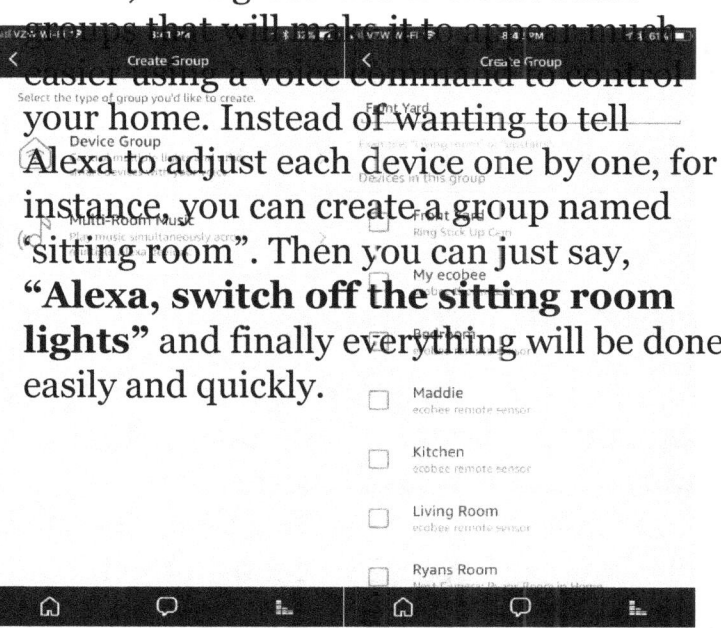

Now, to enable you create group, move to the;

Smart Home, section from the side menu.

Choose, **Groups**

After that select, **Create Group.**

After that, you can now go ahead to customize a name, it's best and safe to use names with 2 to 3 syllables and avoid names that are a little bit stressful. Then you can choose what device to add to the group.

Scenes is another way to customize a smart home setup with the Echo Show. Now, instead of creating scenes, you will be scanning other smart home devices already available for any applicable scenes that has been already created. Hence, this process will give Alexa voice control of already existing scenes from devices like Logitech Harmony Elite and Philips Hue Lightbulbs.

Not able to create your personal scenes, they do provide a good way for convenient and easier control of your existing devices. To search for any available scenes all you have to do is from the side menu just

Go to, **Smart Home**

Scenes.

CHAPTER 6

OTHER FEATURES ON THE AMAZON ECHO SHOW

There isn't any need to panic, as the Echo Show does a lot amazing stuff than controlling a smart home device. Below are rundown of some amazing and interesting features that you can use and explore.

VIDEO CALLING FEATURE

Thanks to the Echo Show touchscreen microphone and 5-megapixel camera, one major feature is the ability to make use of the device for video calling to other people who has the Echo Show or Alexa app. Just like any other Echo products, it does voice messaging and voice calls. The video calling feature is very similar to other services like Apple's FaceTime, but its mode of operations is nice and unique.

To setup and customize the feature, you should go to the Alexa app, then once there choose the conversation portion of the app which is the middle option on the bottom black bar in the app. Its look is like a speech bubble.

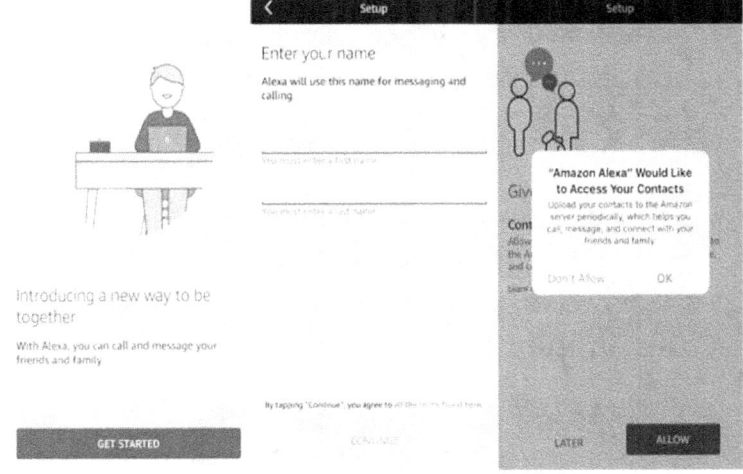

Once confirming your first and last name, you will be asked to verify your phone number via an SMS message. Once that is done, anyone who has your phone number in their contact can use the Alexa app or the Echo device to message or call you.

If need be, you can block individual contacts. To do that, in the conversations portion of the app select the **Contact icon**. Navigate down to the bottom and then press **Block Contacts.**

When you are set to make a call, it can be connected to anyone with the Alexa app, any other Echo device, and it's even possible to a regular number. To do this just say **"Alexa, call (name of the person you would like to call).** Otherwise, you can as well say each digit of the phone number.

You are going to hear a ringing sound, but if the person you are trying to call doesn't answer, Alexa will say the person isn't available. Once you are done with a call, just say **"Alexa, hang up."**

To make a call from the Alexa app, you should choose the **Contacts** icon in the conversation tab and then select the person you want to call. The procedures are similar to what you see when placing a call on your smartphone.

Meanwhile, the mobile number you linked to your Amazon account will be shown on the recipient's caller ID when dialing as a regular phone number.

Note that, when a call comes in, the Echo light icon will turn green and Alexa will tell you who is calling. Here, there are two possible options to command Alexa, just say **"Alexa, Answer"** or **"Alexa, Ignore"**.

Also your smart phone will show a notification that a call is coming in and who the caller is.

DROP-IN FEATURE

The Drop-in feature is available on the Echo line; it is mostly useful on the Echo Show. When someone Drops-in on you, the light bar on your Echo Show will show green, the person with Echo Show or the Alexa app will automatically connect with you and can see and hear anything that is within the range of the device when commanding Alexa for a Drop-in.

Wanting to do a Drop-in, instead of asking Alexa to call someone, you should just say, **"Alexa, drop-in on (contact name)."**

Being aware of the privacy concern, Amazon has made the Drop-in feature completely optional. If you would like a contact to be able to Drop-in, select their details by pressing the contact icon in the **Conversations** portion of the app. Choose the specific contact and then toggle on the Drop-in button.

If dropping in on an Echo Show from another device, or the Alexa app, a frosted view will pop up for the first few seconds. The purpose of this is to give the person time to prepare for the call so that you are not caught off guard. When a Drop-in start, the Echo Show will beep.

ENABLE DO NOT DISTURB FEATURE

The Echo Show Do Not Disturb feature acts like a guard to block Alexa from alerting you of incoming messages and calls. To switch on Do Not Disturb, just say **"Alexa, Do Not Disturb me."** While to switch it off, just say **"Alexa, turn off Do Not Disturb,"** or you can as well click the Do Not Disturb button on the screen of your Echo Show.

VOICE MESSAGES FEATURE

Wanting to send message to another Echo device or the Alexa app, just say **"Alexa, send message to (contact's name).** then you can speak out loud your message. Once you're done talking, the message will send automatically. But if you want to use the app to send a voice message instead,

→ Choose **Conversations**

→ **New conversations**

→ Then the **Contact's name**

Now, you should tap and hold the microphone icon to record your message. After you have finish, remove your finger off the button and it will send.

Note that you can't send an SMS message to any mobile number, as you are limited only to anyone that uses the Alexa app or an Echo device.

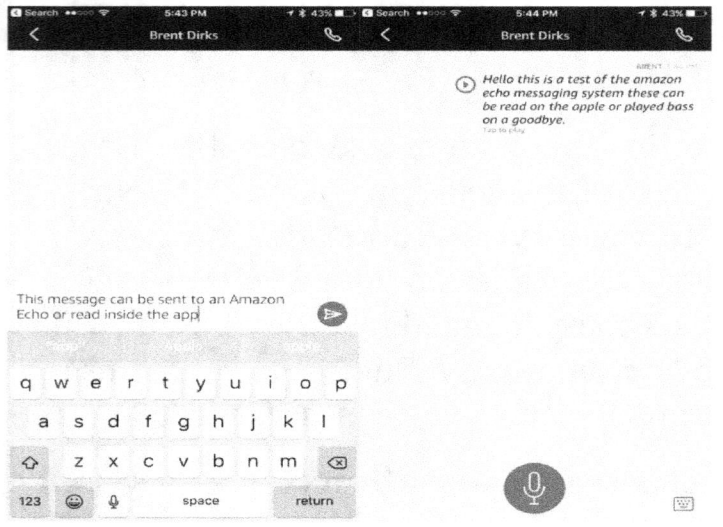

A notification will pop up on the screen after receiving a message, to listen to it just say **"Alexa, play my message"**. While using the app, you should select the particular conversation and press the **"Play"** to listen to your message. Using the app also, simply type in a message and send it to another Echo user. Message can be view on a smartphone or even use the Echo device to say it aloud.

WATCHING MOVIE FEATURE

The Echo Show touchscreen has the ability to play movie from a number of different source. You are able to play purchased content from your Amazon video library, including anything on Amazon Prime Video. Being a subscriber to well know premium channels like Starz, and HBO through Amazon channels, they can be seen on the Echo Show. As well as any movie trailer from IMDb.

Wanting to view or watch a movie or trailer, you should just say **"Alexa, play (movie title).** While watching, you can also give a command to Alexa like, **"Alexa, stop", "pause", "next", "previous", "rewind",** or **"forward".**

SHOPPING ON AMAZON FEATURE

Being the Amazon Echo Show it is possible to shop on Amazon, but it's much easy to shop for anything you can think of on the online retailer using just your voice. If you want to purchase book for instance, just say **"Alexa, order book."** Top rated items will be shown on the screen for that specific search. You can say **"Alexa, show me number one"** to view at the specific item. But when you are ready to make a purchase, you should just say **"Alexa, buy (item name)",** and then follow its instructions.

Using the touchscreen, you can just scroll and make a purchase. However, if you would rather head to the particular brick-and-mortar grocery store, the Echo Show can as well provide you a shopping list.

SETTING A BACKGROUND PHOTO

If need be, to set your own picture as the background photo on your Echo Show's screen, just head to the Alexa app and upload a photo of your choice from your mobile device. From the menu, choose **Settings,** after that tap on your device, and in the Home Screen Background section, from there you should select **Choose a photo.** Alternatively, you are also able to set an album or an entire collection of photos as your background, for as long as they are stored in Prime Photos.

LIGHT BAR DEFINITIONS FEATURE

The bottom of your Echo Show's screen will light up different colors as a visual indicator. Below are the definitions of what each colors mean:

RED: A red bar appears when the microphones and camera of your device are switch off.

ORANGE: An orange bar appears when your device is going through connectivity problems.

PURPLE: A purple bar appears when Do Not Disturb feature is switch on.

BLUE: A blue bar appears on the screen to show that Alexa is processing your request.

CHAPTER 7

TROUBLESHOOTING AMAZON ECHO SHOW PROBLEMS

For much reason that the Echo Show is a very good accessible means of technology that is very easy to use, likely issues act as bumps in the road with daily use. Below are rundown of some issues one might come across when using the voice controllable smart device and means by which you can solve them in a few steps.

"ALEXA" COMMAND NOT ACTIVATING AMAZON ECHO SHOW

One of the most disturbing and common problem is when the **"Alexa"** wake word no longer activate the Echo Show. The seven far field microphones usually do an excellent job listening for the command, even over a loud room. In some cases, the Echo Show doesn't wake up.

Facing such issue, the first thing to check is the Echo Show itself. If it has been put into Do Not Disturb mode unintentionally via a voice command or by simply pressing the button on top of the device, you will see that there is a red LED light close to the camera and the strip on the bottom of the screen is red also. Now, to switch off the Do Not Disturb mode just press the button on top of the Echo Show.

Next thing to do is, check if the Echo Show is close to an audio interference like a television, fan, or a radio, if so the microphones might be unable to hear the command. The best and suitable solution is moving the device to a very comfortable location in your house.

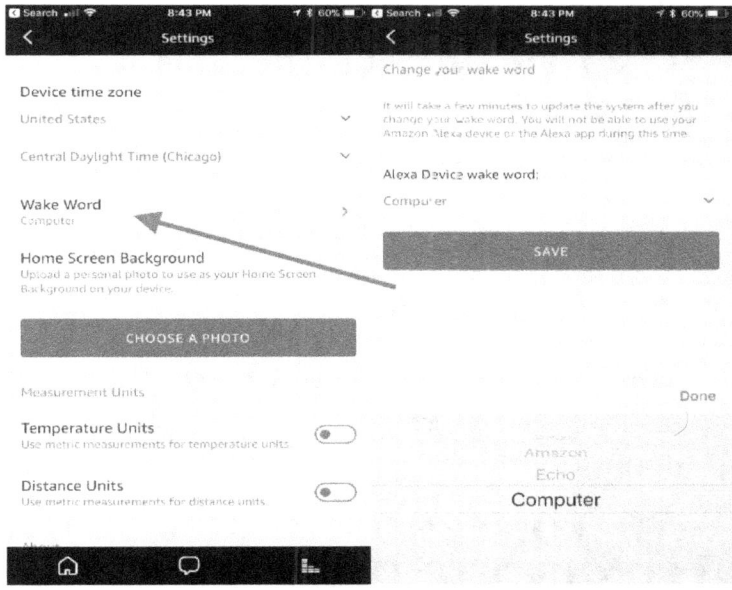

Also, ensure that the wake word hasn't been changed to another option like "Computer", "Amazon", or "Echo". To correct this, go to **Settings** in the companion app. Look for the Echo Show and then move down to **Wake Word.**

Another thing to check is the Voice Training feature in the Alexa app. It helps to improve the speech recognition of the Echo Show. To fix this, from the side menu in the app,

→ Choose **Settings**

→ **Voice Training**

→ **Start Session.**

You should speak the phrase on the screen and press **"Next"** until the session is fully complete.

AMAZON ECHO SHOW UNABLE TO CONNECT TO THE INTERNET

This shouldn't be seen as a surprise, but for the device to work effectively, a full time internet connection is needed. When the device is not connected to the internet, you will notice an icon on the screen of a

Wi-Fi signal with line that passes through it. The light bar changes to orange.

You should ensure to turn on by moving your hand downwards from the top of the Echo Shown screen and then you should select;

→ Settings

→ Wi-Fi.

In some devices, the first thing to do is just to power cycle the Echo Show for a minute and hope it regain an internet connection. But if not try restarting your Wi-Fi router and also your modem.

If still that doesn't help, then you should bring the router closer to the Echo Show or vice versa. Hopefully taking these actions will solve the problems. Or if need be, you should try upgrading your Wi-Fi router or even purchase a mesh Wi-Fi system.

"ALEXA" ACCIDENTAL OR RANDOM RESPOND ON THE AMAZON ECHO SHOW

Another common issue one might face is the Echo Show might respond even if you didn't say a word or give a command. At this point, the best solution is moving the Echo Show away from any possible source of audio interference, be it a TV or radio. Echo Show microphones are highly sensitive, that even the word **"Alexa"** mention on a TV or radio could cause an accidental or random response.

Therefore, try changing the wake word and see if that helps solve the issue. This step is very helpful if you are living in a house with someone bearing a name that is similar to Alexa.

OTHER ECHO DEVICES IN YOUR HOUSE RESPOND AT ONCE

If the Echo Show isn't the only Alexa device in your house, this may result to a disturbing issue. Immediately you say the **"Alexa"** wake word, more than one device will respond. All Echo device from Amazon features a special technology called Echo Spatial Perception. This feature is supposed to provide limits in a way that only the closest Echo device responds to a given command. But for the most parts, the ESP works nicely, apart from other Alexa enabled devices that are not from Amazon (like the Ecobee4 Smart Thermostat) they don't offer the same ESP technology.

The first thing to do to give a possible solution to this issue, is separating the Echo Show further away from any other Alexa devices.

But, if that doesn't fix the problem, you should try changing your Echo Show's wake word to any of the available options.

SHOW AND TELL

In a nutshell, the Echo Show takes all the features that make the real connected speaker line so widely known and added a new world where there are possibilities with the Echo Show 7-inch touchscreen.

After going through this guide and reading it, hopefully you have gained a good understanding and being enlighten of the Echo Show and how you can explore and make use of its amazing features more interesting and enjoyable.

THE END